Bimal Kumar Jha
Anjana Deva

Role of Boron in Steel : An Overview

Bimal Kumar Jha
Anjana Deva

Role of Boron in Steel : An Overview

LAP LAMBERT Academic Publishing

Impressum / Imprint

Bibliografische Information der Deutschen Nationalbibliothek: Die Deutsche Nationalbibliothek verzeichnet diese Publikation in der Deutschen Nationalbibliografie; detaillierte bibliografische Daten sind im Internet über http://dnb.d-nb.de abrufbar.

Alle in diesem Buch genannten Marken und Produktnamen unterliegen warenzeichen-, marken- oder patentrechtlichem Schutz bzw. sind Warenzeichen oder eingetragene Warenzeichen der jeweiligen Inhaber. Die Wiedergabe von Marken, Produktnamen, Gebrauchsnamen, Handelsnamen, Warenbezeichnungen u.s.w. in diesem Werk berechtigt auch ohne besondere Kennzeichnung nicht zu der Annahme, dass solche Namen im Sinne der Warenzeichen- und Markenschutzgesetzgebung als frei zu betrachten wären und daher von jedermann benutzt werden dürften.

Bibliographic information published by the Deutsche Nationalbibliothek: The Deutsche Nationalbibliothek lists this publication in the Deutsche Nationalbibliografie; detailed bibliographic data are available in the Internet at http://dnb.d-nb.de.

Any brand names and product names mentioned in this book are subject to trademark, brand or patent protection and are trademarks or registered trademarks of their respective holders. The use of brand names, product names, common names, trade names, product descriptions etc. even without a particular marking in this works is in no way to be construed to mean that such names may be regarded as unrestricted in respect of trademark and brand protection legislation and could thus be used by anyone.

Coverbild / Cover image: www.ingimage.com

Verlag / Publisher:
LAP LAMBERT Academic Publishing
ist ein Imprint der / is a trademark of
AV Akademikerverlag GmbH & Co. KG
Heinrich-Böcking-Str. 6-8, 66121 Saarbrücken, Deutschland / Germany
Email: info@lap-publishing.com

Herstellung: siehe letzte Seite /
Printed at: see last page
ISBN: 978-3-659-37138-7

Zugl. / Approved by: Ranchi, Ranchi University, 2012

Copyright © 2013 AV Akademikerverlag GmbH & Co. KG
Alle Rechte vorbehalten. / All rights reserved. Saarbrücken 2013

TABLE OF CONTENT

SL. NO.	CONTENT	PAGE NO.
1.	Introduction	1
2.	Boron in Hardenable Steels	2
2.1.	Grain Boundary Segregation of Boron	6
2.2.	Boron-Nitrogen Interactions in Al-killed Steel	7
2.3.	Grain Size Control & Grain Coarsening Temperature	12
3.	Boron for Formable Steel	16
3.1.	Hot Rolled Steel	16
3.1.1.	Hot Band Microstructure	21
3.1.2.	Thin Gauge HR	22
3.1.2.1.	Austenitic Rolling	22
3.1.2.2.	Ferritic Rolling	23
3.2.	Cold Rolled Steel	27
3.2.1.	Continuous Annealed Steel	27
3.2.2.	Batch Annealed Steel	32
3.2.3	IF steel	35

Role of Boron in Steel: An Overview

Anjana Deva and B K Jha

Research and Development Centre for Iron and Steel,
Steel authority of India Limited, Ranchi, India
* Email: bkjha@sail-rdcis.com

Abstract

Out of all the elements boron is unique one, which can increase or decrease the hardness of steel depending on presence of Titanium or Zirconium, nitrogen content and process condition i.e austenitising temperature, cooling rate etc. It has been well established that addition of a small amount of boron (10-30 ppm) remarkably increases the hardenability of low alloy steels. This beneficial effect is attributed to the grain boundary segregation of boron, which retards the transformation of austenite to ferrite by reducing the grain boundary energy. However, boron effect is entirely different in low carbon alloyed steel processed under controlled condition. This review paper discusses the intricacy of boron addition in low carbon unalloyed steel to medium carbon low alloyed steel. Boron has been found to act as softener in the former and hardener for the latter.

1. Introduction

Significant progress has been made towards the development of new steel products with improved attributes by addition of micro-alloying elements in the past [1-3]. The properties attained in the final product depend on presence of small additions of niobium (Nb), vanadium (V), titanium (Ti) or boron (B), which by precipitating in the form of carbide and carbo-nitride, can give precipitation hardening and / or inhibition of grain growth. Out of these elements boron is a unique one, which can increase or decrease the hardness of steel depending on presence of titanium or zirconium (Zr), nitrogen (N) content and process conditions i.e. austenitising temperature, cooling rate etc. [4-17].

2. Boron in Hardenable Steels

Boron has aroused interest of many materials' scientists due to its hardenability enhancing effect, when used as an alloying element in steels [4-11]. The effect of boron and the need to control its location and chemical state are known since 1940's [4]. Early attempts to commercially produce heat treatable steels resulted in variable properties due to difficulties in ensuring that boron was present in desired form, e.g. soluble boron in case of steels where boron has to enhance hardenability. Subsequently, the importance of protecting boron using strong nitride formers such as titanium was appreciated [6]. However, problems can be encountered in obtaining a consistent effect as can be seen from wider hardenability bands that are specified for boron treated steels compared to their Cr / Mo / Ni equivalents. On the other hand, similar problems can be encountered in areas where an unprotected boron addition is made to modify other properties, for example, ductility or formability in low carbon steels.

During the last two decades, the consensus has been that only a small amount of dissolved boron would be effective to increase the hardenability of steel [10]. The suggested mechanism for the same is based upon the assumption that boron in elemental form segregates at the prior austenite grain boundaries suppressing the ferrite reaction and thus improving the hardenability. Several authors have applied different testing methods to prove the existence of boron in elemental form in austenite grain boundaries. Grossmann [18] observed that in a 0.6 wt.% carbon steel, hardenability increased to a maximum with addition of boron up to

25 ppm and then decreased with larger additions. Other investigators likewise [19] reported maxima in the hardenability at about 30 ppm. Craft [20], however, found that in commercial open hearth heat of medium carbon steel, the hardnability increased linearly with boron up to 100 ppm. These observations suggest that the effectiveness of boron in increasing hardenability depend on the form of boron retained in the steel. These forms are being influenced by the presence of other elements like oxygen, nitrogen, titanium, zirconium etc. The importance of achieving adequate de-oxidation prior to the addition of boron has been emphasized by many investigators [21-23]. Many authors [24-26] have investigated the role of nitrogen and have shown that nitrogen also interacts with boron and reduces its effect on hardenability.

Carbon is yet another element, which has a pronounced effect on boron-induced hardenability in steel. It has been shown [22,27] that hardenability effect of boron diminishes with increasing carbon content and becomes almost negligible at the eutectoid composition. Kapadia et.al [28] has studied the influence of nitrogen, titanium and zirconium on the boron-induced hardenability in constructional alloy steel and has brought out a physically meaningful correlation between the observed hardenability effects of boron, as explained by the following expression.

$$B_{eff} = [\,B - \{(N - 0.002) - Ti/5 - Zr/15\}_{\geq 0}\,]_{\geq 0} \qquad (Eq.\ 1)$$

In the expression, the chemical symbols represent the total weight percent of the respective elements contained in the steel whereas B_{eff} is the effective boron participating in enhancement of hardenability. According to the above equation, the boron combined with nitrogen is equivalent to the weight percentage of nitrogen available after combining with aluminum, silicon and vanadium, which together account for a fixed amount of 0.002 wt.% N. The resulting relationship between observed hardenability and effective boron is shown in Fig. 1. The form of the observed relationship between hardenability and effective boron (Eq. 1) is consistent with result of another study by Grange and coworkers [27,29]. They concluded that boron contents above the optimum amount tends to promote the precipitation of boron-rich grain-boundary-constituent, which depletes the austenite grain boundaries of boron atoms and results in an overall loss of effectiveness.

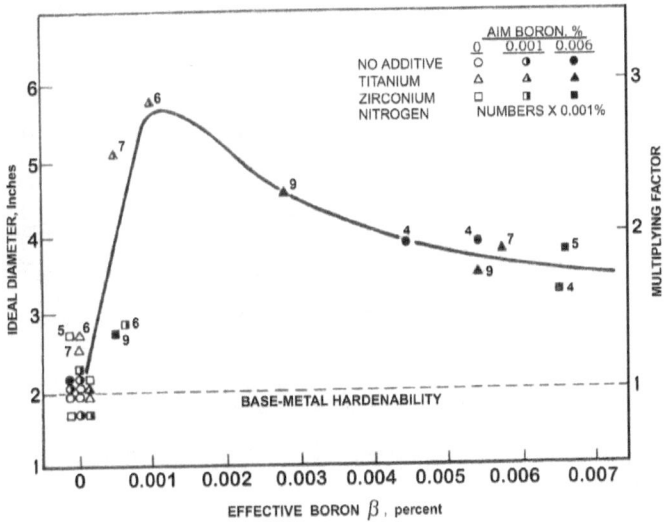

Fig.1   Effect of effective – boron content on hardenability of the steels

To protect boron from nitrogen, titanium, a potent nitride-forming element, is frequently used. Titanium nitride (TiN) forms in the liquid state prior to solidification and is a very stable compound in the solid state. Hence, it does not dissociate during heat treatments and when added in proper quantity, it efficiently blocks nitrogen.

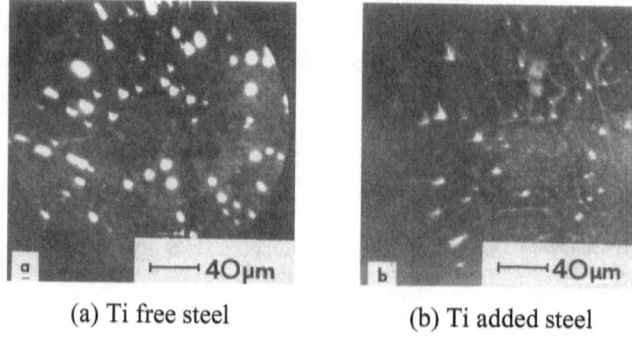

(a) Ti free steel   (b) Ti added steel

Fig. 2   Ion micrographs showing protective effect of Ti

Two ion micrographs (Fig. 2 a&b) provide a clear illustration of protective effect of Ti; after a standard step quenching experiment, the steel without titanium showed no evidence of grain-boundary-boron, whereas the steel containing titanium exhibited boron enrichment at the grain boundary through precipitation of $M_{23}(B,C)_6$, as observed in low nitrogen boron-steels [10].

Hardenability of boron-steel is affected by the precipitation behaviour of boron. In the work of Watanabe and Ohtani [30] with different kinds of precipitates of boron, which change their state depending on the heating temperature before rolling and quenching temperature during heat treatment of Al-B treated steels, they found that when the steels are furnace cooled (cooling rate of 0.67 °C / min.) after hot rolling, AlN grows radially from BN which precipitated before the nucleation of AlN. AlN formed on BN during furnace cooling is dendritic. Al, B and N dissolve into solid solution during reheating of slab at 1300 °C. On subsequent hot rolling and post cooling, aided by high diffusivity of B and N over Al, concentration of B and N atoms increases at grain boundaries to the extent that even solubility product is exceeded while Al atoms remain distributed homogeneously and do not prevent BN to precipitate. Thus, due to high diffusion velocity, BN precipitates on austenite grain boundaries. Under the situation, the effect of boron on the hardenability almost disappears. This situation corresponds to non equilibrium state of Al-B-N-Fe system. Habu et.al [31] showed by calculating B content in solid solution in the equilibrium state of Al-B-N-Fe system, that Al protects B from precipitating as BN by fixing N as AlN, which ensures presence of B in solid solution, needed for hardenability.

Boron hardenability mechanisms have been reviewed by Morral and Cameran [7]. With the possibility of boron concentration reaching significant levels, a number of mechanisms for retarding ferrite nucleation have been discussed by the authors with due consideration of reduction in austenite grain boundary energy, reduction in diffusivity, reduction in number of sites and nucleation of ferrite on borocarbides. It has been concluded in the review that the proposed boron hardenability mechanisms are remarkably similar, except for details of how the ferrite 'C' curve changes. Their prediction regarding grain size, austenitising temperature and borocarbide concentration are qualitatively the same. The reason is that all the mechanisms depend directly or indirectly upon the concentration of boron absorbed in the austenite grain boundaries during austenitising. These theories predict that achieving the maximum concentration of free boron (presumably by avoiding boron interaction with oxygen or nitrogen) is the key to optimizing the boron effect.

2.1. Grain Boundary Segregation of Boron

The topic of boron segregation to austenite grain boundaries is still a domain of controversial experimental evidence. There is little doubt that boron can easily migrate to austenite grain boundaries, and the mere fact of the occurrence of grain boundary precipitation of borocarbides during cooling is good evidence of this behaviour. The basic question is the existence of equilibrium segregation of boron at the grain boundaries in the temperature range of interest for austenite decomposition (below 800 °C).

Several authors [32,33] have argued on the basis of alphagraphy experiments wherein boron segregation was unambiguously observed prior to the onset of precipitation of borocarbides. However, it has also been shown that alphagraphy is not able to differentiate between segregation and precipitation [32,33]. The results presented by Ohmori and Yamanaka [8] brought fresh evidence of grain boundary enrichment, through high sensitivity ion microprobe analysis. However, it is likely that the enrichment found, also covers grain boundary segregation and precipitation of borocarbides, as admitted by the authors. The use of Auger electron spectroscopy has also been rather deceptive, due to a) the difficulties in obtaining intergranular rupture surfaces, and b) interference between boron and chlorine peaks.

It is widely known that nitrogen is a "poison" [6] as far as the boron hardenability effect is concerned. In steels where an adequate protection of boron by nitride forming elements is not there, boron combines with nitrogen as BN. This compound has been identified by X-ray diffraction on residues obtained by chemical extraction ("insoluble" boron in dilute sulphuric acid) as well as selected area diffraction. Because boron has affinity for oxygen comparable to that of silicon [34], boron steels must be thoroughly killed with aluminium in order to prevent the reaction of boron with oxygen [35]. Boron oxide, B_2O_3, is much more stable than the boron nitride, as indicated by the standard free energy of formation at 980 °C : –235 kcal per mol of B_2O_3 (liquid) and –28 kcal per mol of BN (crystal) [36]. It is evident that the oxygen potential must be kept low in presence of BN in order to prevent its decomposition and subsequent formation of B_2O_3. Limited data are available, however, concerning the distribution of BN in the microstructure. In the work of Watanabe and

Ohtani [37], boron nitride precipitates were found as intergranular and intragranular particles depending on the thermo mechanical history of the specimen.

2.2. Boron-Nitrogen Interactions in Al-killed Steel

BN is thermodynamically less stable than AlN. However, in Al-killed steel, where aluminium additions are made for killing during steelmaking [38], BN is less soluble in austenite than AlN. This apparent anomaly is considered to arise because of the high rate of diffusion of boron in austenite, which is nearly same as that of nitrogen in austenite [39]. Boron appears to act interstially in austenite [40]. Under appropriate heat treatment conditions, therefore, BN precipitates in preference to AlN.

Fig. 3 illustrates an isothermally transformed low carbon B-Al steel. More than 50 % of N precipitates as the equilibrium quantity of BN after 1 minute

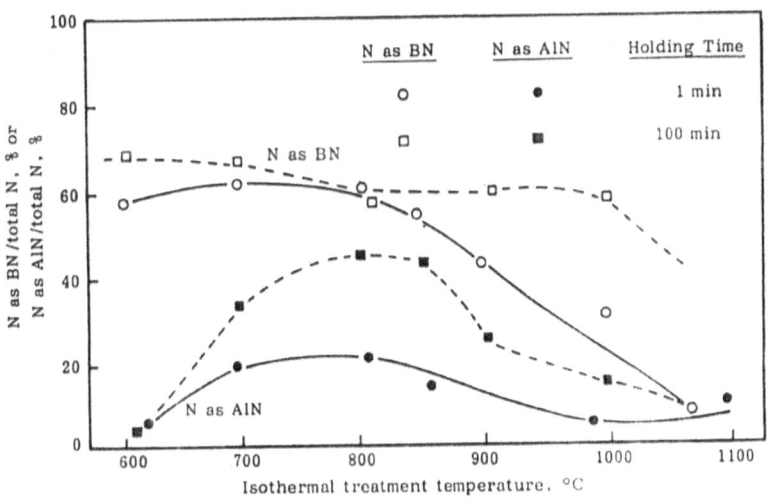

Fig. 3   Precipitation of BN and AlN in B-Al steel

at temperatures less than 900 °C. However, even after 100 minutes at temperature, less than 900 °C, 40 % of N precipitates as the equilibrium quantity of AlN [41]. It is evident that the maximum rate of precipitation of AlN coincides with γ to α transformation. The precipitation of BN is so rapid that it has been detected even in as quenched condition [42]. Even aftervery short hot rolling cycles, BN precipitates at a rate close to equilibrium. Engl and Drews [39] showed that if B/N $\geq$ 1, BN precipitation can occur even with coiling temperature as low as 520 °C, in marked contrast to the effect of coiling temperature on AlN precipitation.

The competitive precipitation of AlN and BN has been studied by several authors [43,44]. It has been found that AlN can form at the expense of BN during equilibrium treatments. In low alloy steels, various formulas [43,44] have been given for the BN and AlN solubility products (Eq 2 & 3) as shown below:

Log_{10} [Al][N] = $-$ 7400 / T + 1.95 (Eq. 2)
Log_{10} [B][N] = $-$ 13970 / T + 5.24 (Eq. 3)

In 4025 type low alloy steels, with very low Al contents, Eq. 4 obtained from analysis of "soluble" boron by sulphuric acid method has been found more appropriate [45].

Log_{10} [B][N] = $-$ 6700 / T + 0.20 (Eq. 4)

Yamanka and Ohmori [45] have examined the effect of prolonged reaustenitising treatment on the relative stabilities of BN and AlN in previously solution treated 0.2 C-Mn-0.044 Al-B-N steels. They found that the non-equilibrium BN always precipitates before AlN. Non-equilibrium BN may even be present in as quenched condition as per Habu et.al [42]. However, with increasing austenitising time, reversion of BN occurs in lieu of eventual equilibrium quantities of AlN. Both Habu et.al [46], and Engl and Drews [39] have used equilibrium thermodynamics to calculate relative stabilities of BN and AlN in steels.

Kapadia [38] has criticized the above method, pointing out that such approaches assume that nitrides are precipitated in equilibrium quantity, when it is clear that generally it is not the case. Kapadia has also questioned the accuracy of current analytical techniques for determination of solute boron. Yamanka and Ohmori [45] extracted nitrides by the Beeghly method subsequently separating the AlN from BN in a solution of hydrochloric acid. In view of the known limitations of these separation techniques, such reservations appear to be justified [38].

Takahashi et.al [47] has studied the precipitation behaviour of BN and AlN during hot rolling. In laboratory simulation, about one minute or so of holding in the austenite range below 1000 °C, the amount of BN precipitate was more than 50 % of the equilibrium amount as calculated using the equation of RW Fountain [44]. AlN, on the other hand, precipitated only less than 30% of the equilibrium amount computed with the help of the equation of WC Leslie [43] even after precipitation treatment for as long as 100 minutes. Below 900 °C, rate of precipitation increases rapidly. This indicates clearly that BN precipitates preferentially in austenitic temperature range even when boron and aluminium both are present. It is assumed from the precipitation behaviour of nitride during hot rolling that BN could precipitate very rapidly in the austenitic temperature range, where as AlN could hardly precipitate in that temperature range. However, in ferritic temperature range, AlN precipitates with relative ease. It may be noted that BN precipitation rate is not affected by coiling temperature either and virtually all the nitrogen contained in the steel combines with boron as BN during hot rolling. These findings are in good agreement with the experimental results **[47]** concerning the precipitation behavior of nitrides. Lower slab reheating temperature resulted in higher BN precipitation. Complete precipitation of BN was not found in actual hot rolled steel and it varied from 60-80 % (N as BN / total N).

The solute B and N contents in the Al-B-N-Fe system have been calculated using mass balance and solubility data **[49]**. Neglecting presence of the small amount (0.001-0.002 wt.%) of Ti, the nitrogen in solution [N] can be expressed in terms of AlN and BN **[43,44,49,50]**.

The accuracy of solubility model depends largely on reliability of the solubility data particularly that of Al/N which varies considerably in the literature **[51]**. The calculated equilibrium [B] and [N] contents for the various low carbon steels (C: 0.03-0.05 wt.%) and varying B/N ratio have been plotted and typical plots are shown in Fig. 4.

In steels with low B/N ratio (0.1 to 0.15), BN precipitation is generally predicted to start between 950 °C and 1000 °C, which is typical temperature range of finish rolling. In the steel with B/N 0.57 to 0.93, BN precipitation is predicted to begin in between 1050 to 1150 °C which is typical temperature range of rough rolling. The prediction that boron is completely in solution after reheating to 1150 °C in all boron added steels has been confirmed by Transmission Electron Microscopy (TEM), where no BN particles were found **[49]**.

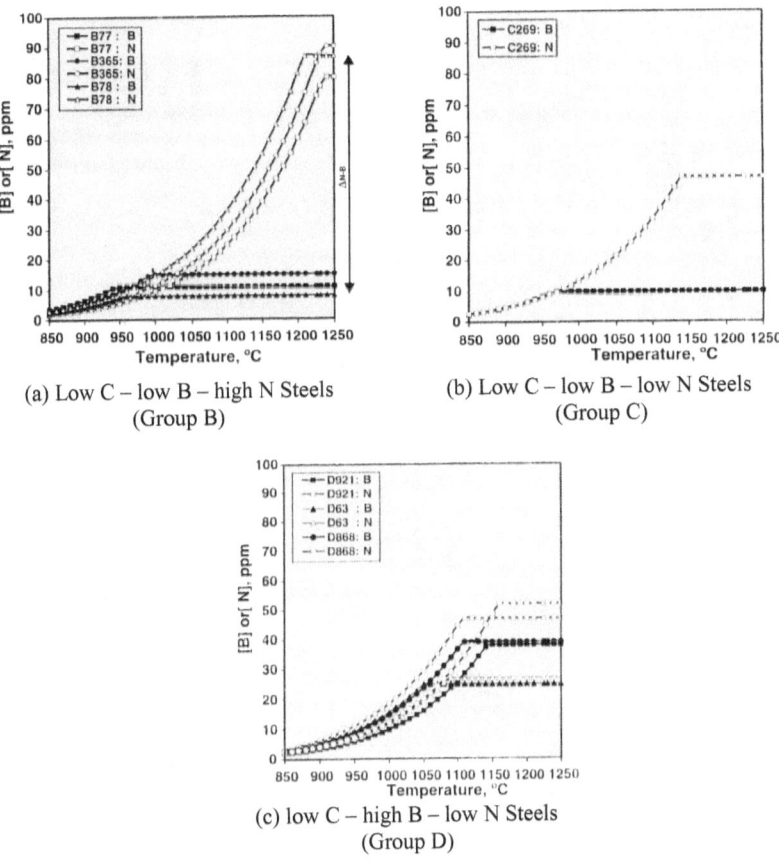

Fig. 4   Calculated Equilibrium [B] and [N] content [49]

2.3. Grain Size Control & Grain Coarsening Temperature

Grain size is of vital importance in controlling the properties of steel. In transformable steels, for fixed cooling conditions, the ferrite grain size is directly dependent on the prior austenite grain size. When plain carbon steel is soaked at progressively higher austenitising temperatures, grain coarsening occurs continuously, causing a gradual increase in size of the relatively uniform equiaxed austenite grains. The driving force for grain growth arises from the reduction of grain boundary area and hence grain boundary energy, which accompanies grain growth [52,53]. The austenite grain growth characteristics of Al-killed low carbon steel are depicted in Fig. 5.

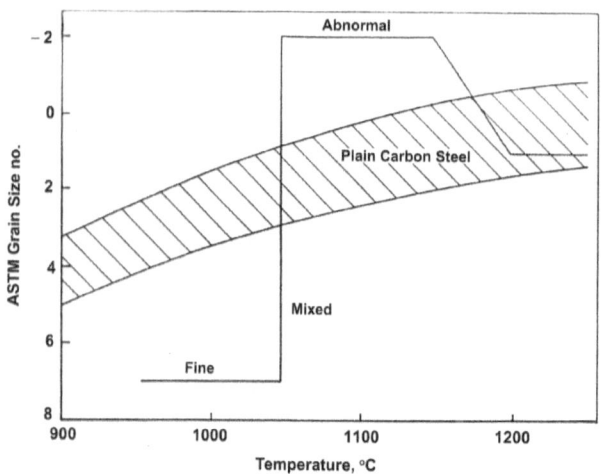

Fig. 5   Austenite grain growth characteristics of Al-treated low C steel

It is perceived that boron does not contribute to grain refinement of Al-killed steel. However, recent study by Bank et.al [49] is of great significance as far as effect of boron on austenite grain growth is concerned. The authors have carried out study on low carbon Al-killed steel with varying B/N ratio. The austenite grain sizes at various reheating temperatures are shown in Fig. 6 for steel group A (which has low carbon – low nitrogen), steel group B (which contains low carbon – low boron – high nitrogen), steel group C (having low carbon – low boron – low nitrogen), and steel group D (having low carbon – high boron –low nitrogen contents).

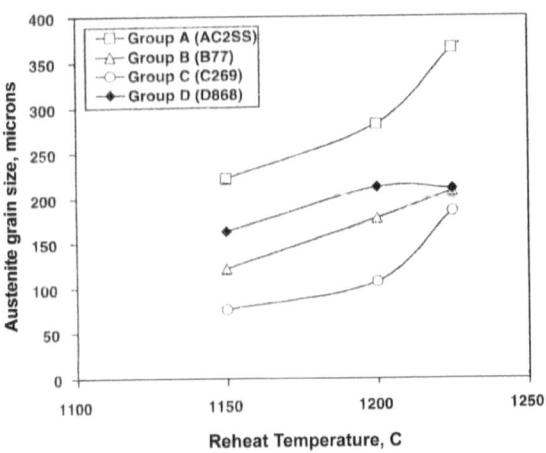

Fig. 6   Austenite grain medians for Groups A – D [49]

Steel group A, which is boron free, has larger Dγ (austenite grain size) than all other boron added steels at a given temperature. For reheating temperature (RHT) of 1150 °C, steel group B with B/N of 0.4, though having mean grain diameter 114 μm, had a bimodal grain size distribution with median of 30 and 210 μm indicating partial coarsening. Steel group D with B/N of 0.75 had a more uniform coarse grain size (160 μm) at this RHT. At a RHT of 1225 °C, group B (B/N = 0.4), group C (B/N = 0.21) and group D (B/N = 0.75) had Dγ between 205 and 220 μm, which is smaller than the Dγ of 363 μm for boron free steel (i.e. Steel group A). No BN precipitate could be detected in steels of group B with B/N = 0.4 and group D with B/N = 0.75 after reheating at 1150 or 1225 °C indicating complete dissolution of boron. At RHT of 1225 °C, no AlN precipitate was found, implying complete dissolution of AlN precipitates. However, above 1150 °C a few coarse AlN precipitates associated with large (> 1000 nm) MnS particles were found. The reason for less extent of grain coarsening in boron added steel has been attributed to solute drag effect of boron. Stampf and Bank [54], also, report that segregated boron on moving grain boundaries retard the mobility of these boundaries through solute drag thereby retarding grain growth of austenite.

In the study, Masatosha Sudo [91] has concluded that there is no apparent difference in austenite grain size among the steels having B/N of 0.7, 1.5 and 2.0 when soaked at 1150 for 30 minutes (average grain size of 140 – 150 µm). However, peak ferrite grain size of hot band at B/N of 1.5 has been attributed to the decrease in nucleation sites during transformation from austenite to ferrite. Boron has been found to increase the temperature of no recrystallisation (Tnr) of both hot and cold rolled steel [55]. Bain et.al [56] in their work have shown that Tnr is higher in Nb-B steel than in Nb steel. The phenomenon has been explained in terms of influence of boron on the precipitation kinetics of Nb(C,N). Further, it has been suggested that contribution may also be made by the non equilibrium segregation of boron to austenite-grain boundaries [57-61]. In cold rolled steel, recrystallisation start and finishing temperatures rise drastically with increasing B content up to 5 ppm and thereafter (i.e. boron content more than 5 ppm) the rise in recrystallisation temperature becomes gradual.

Since the nucleation of ferrite occurs primarily on grain boundaries, the grain size is an important variable in hardenability. It has been reported that the addition of boron tends to increase austenite grain size. Ohmari and Yamanaka [8] mentioned that the observation can be an alloy-impurity boron effect. However, in practice boron steels often have smaller grain sizes than similar boron free steels because elements which can act as grain refiners are added with boron to protect it from oxygen and nitrogen.

A quantitative estimate of the grain size effect can be written by assuming that ferrite nucleation occurs primarily on grain faces [7], and therefore can be expressed as Eq 5 :

$$nv = nASv \quad (Eq. 5)$$

where, Sv is the area of austenite grain boundaries per unit volume and nA is the number of nucleation sites per unit grain boundary area.

3. Boron for Formable Steel

Conventionally, boron has been added for improving the hardenability of low-alloyed quenched and tempered steels as discussed in previous sections. However, the advancement of steel making process made it possible to control boron content in steel, even at ppm levels and thereby controlling B/N ratio. Addition of boron in desired amount with controlled processing of steel can lead to improvement in formability of low carbon steel, which has drawn considerable attention in recent past. In the succeeding sections, the effect of boron addition on microstructure and properties of hot rolled and cold rolled low carbon Al-killed steel shall be discussed.

3.1. Hot Rolled Steel

Hot rolled sheet steels are normally produced to satisfy specifications such as yield strength, ultimate tensile strength and total elongation. Typically, as yield strength (or yield strength - to – ultimate tensile strength ratio) increases, ductility decreases. Total elongation is also a general measure of formability; higher elongations generally provide better formability. However, the elongation value is inversely proportional to the thickness of hot rolled sheets.

The formability of hot rolled steel is characterized by lower yield strength, low YS/UTS ratio, high elongation, high strain hardening exponent (n) and low strain aging index (SAI).

The cold formability of mild, unalloyed hot rolled steel is seriously impaired by nitrogen dissolved in ferrite [62,63]. Solid solution hardening and above all aging by nitrogen are responsible for this phenomenon, which can be counteracted by optimized temperature management in the hot strip mill in case of Al-killed steel. However, with the high coiling temperature necessary for nitrogen fixation, a limited degree of uniformity of the mechanical properties over the length of strip as well as unfavourable pickling behaviour is expected. Fixing of nitrogen by Ti or B provides a remedy here. In the study by Muschenborn et.al [64], it has been shown that titanium or boron addition improves the formability of hot rolled steel as a result of nitrogen fixing. Further, they have shown that titanium addition results in higher yield strength and less favourable total elongation with the same level of nitrogen fixing. The reason for this may be attributed to grain refinement potential of Ti.

Although titanium has been extensively used to tie-up nitrogen, potential of boron for this purpose has not been fully exploited. YR Cho and SI Kim [65] in their recent work have studied the effect of boron addition on the microstructure and mechanical properties of as hot rolled unalloyed low carbon (0.02-0.04 wt.% C) steels. In their study, the yield and tensile strengths of boron free low carbon steel have been found to be constant with increasing finish rolling temperature. However, the strength, especially the yield strength of boron added steel decreased with increase in rolling temperature. This dependence of mechanical properties on rolling temperature for boron added low carbon steel has been attributed to the different microstructural evolution and precipitation behaviour of the steel. Ferrite grains of boron added steels became coarse and irregular at higher rolling temperatures. However, in the case of the boron free steel, grain coarsening has not been observed (Fig. 7).

In addition, as rolling temperature increased, fine cementite formation has been found to be more common in boron added than in boron free low carbon steels. In boron free low carbon steels (Fig. 8), the volume fraction of cementite increased with decrease in rolling temperature [65]. On the

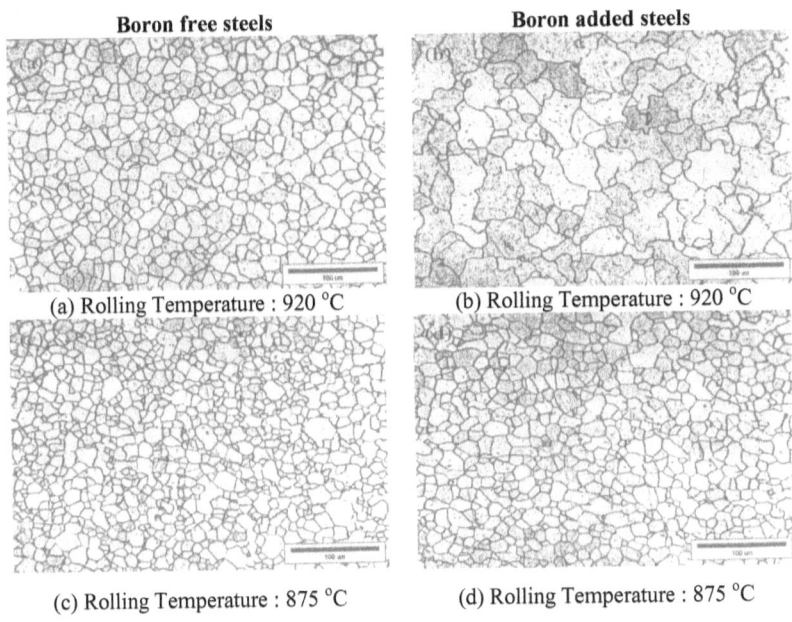

Fig. 7 Optical microstructures of hot rolled sheets for boron-free (a & c) and boron added (b & d) low carbon steels.

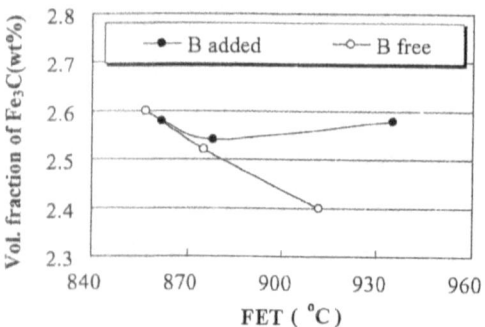

Fig. 8 Volume fraction of cementite obtained from microstructural analysis at various rolling temperatures in low carbon steels

other hand, in boron added low carbon steel, the volume fraction of cementite increased drastically at higher rolling temperature.

YR Cho and SI Kim [65] have concluded that boron improves the elongation of steel due to reduction of solute nitrogen and carbon that is caused by precipitation of boron nitride / carbide as well as the higher density of cementite. The size and density of BN precipitates have been found to increase with decreasing coiling temperature. In case of boron added low carbon steel, the yield strength showed the lowest value at 20 ppm of nitrogen content. (Fig. 9).

Above 20 ppm of nitrogen, increase in the content of nitrogen increases the strength of boron added low carbon steel. However, the yield strength of boron free low carbon steel increases with increasing nitrogen content monotonically. The relationship between yield strength and nitrogen content

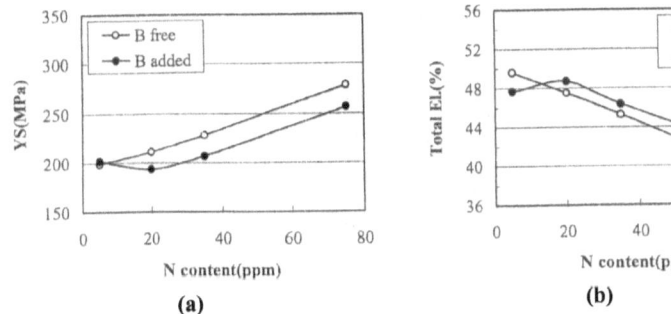

Fig. 9 Effect of B and N addition on the mechanical properties of low carbon steel : (a) Yield strength and (b) total elongation

can be explained by the B/N ratio. When the nitrogen content surpasses the B/N ratio of unity, excess solute nitrogen in the ferrite matrix increases the strength and deteriorates the ductility of steel.

It is mentioned in their work [65] that distribution of precipitates increased with decreasing finish rolling temperature in boron added steel, whereas in boron free steel, the MnS and AlN were mainly investigated, and their density of precipitates was not found to be dependent on rolling temperature. It has also been observed that the volume fraction of precipitates was higher than that of boron free steel because BN precipitates grew rapidly on CuS and MnS.

Messien and Leroy [66] in their study on hot rolled low Al-B steel have shown that low nitrogen (Nsol) in boron added steel are achieved owing to an extensive BN precipitation as the densities of the precipitates appear to be independent of coiling temperature. Very fine AlN precipitation has been observed and its density depended on the coiling temperature, particularly coiling at 650 °C and to a lesser extent on the Al-content of steel.

3.1.1 Hot Band Microstructure

Pradhan [67] in his work on effect of boron on mechanical properties of hot rolled steel has explained the reason for coarser ferrite in boron steel. Precipitation of boron nitride occurs in the austenite temperature regime and is essentially completed at temperature below 900 °C [47], whereas AlN can only precipitate in the ferrite temperature regime. As such, compared to AlN precipitate particles, the BN precipitates are much coarser. In plain Al-killed steel, the fine AlN particles result in relatively fine ferrite grain size (16 μm) even at high coiling temperature of 675 °C. Addition of boron leads to partial or complete substitution of fine AlN with coarser BN, resulting in coarser ferrite grain size. When B corresponding to the stochiometric ratio (B/N 0.8) was excluded, the excess B caused ferrite grain refinement, probably due to one of the two reasons (a) excess B lowers the ferrite transformation temperature, or (b) it forms fine iron boro-carbides [47,68,69]. At a given B/N ratio, decreasing carbon content led to coarsening of ferrite grain size and a consequent decrease in hot band strength level (Fig. 10), numbers in bracket indicate ferrite grain size at coiling temperature of 675 °C.

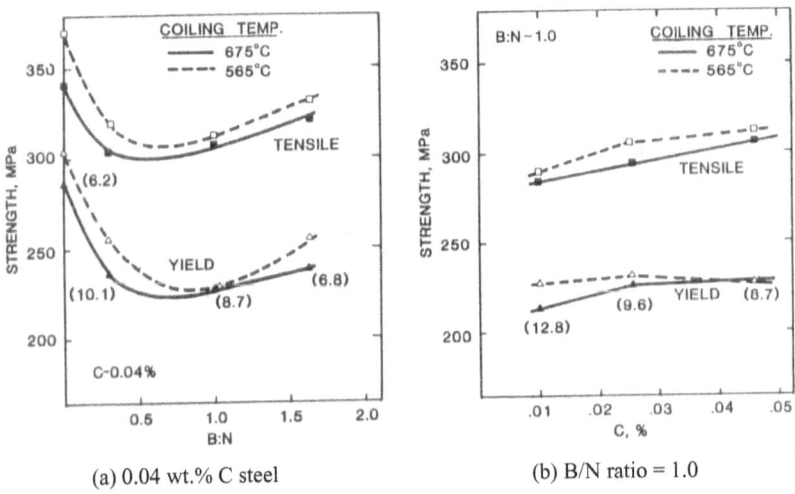

(a) 0.04 wt.% C steel

(b) B/N ratio = 1.0

Fig. 10 Effect of B/N ratio for (a) 0.04wt.% C steel and (b) C content for B/N ratio of 1.0 on yield and tensile strength

3.1.2 Thin Gauge HR

3.1.2.1 Austenitic Rolling

Cold rolled thin gauge steel sheet is normally rolled from low carbon hot rolled steel of relatively thin gauge. To keep the roll force within limits even during severe rolling, the hot rolled (HR) coils must have low hardness, and also the pick-up of hardness of the stock during rolling must be low (i.e. HR coil should not work harden rapidly).

Boron reduces phase transformation in the steel, which enhances possibility of hot rolling above Ar_3 temperature, without loss of ductility caused by mixed microstructure during the production of thin gauged steels. Boron, accordingly, has been found [70-74] to be highly suitable for thin gauge cold rolling and forming application.

Boron treatment does not lead to increase in the rolling load in hot strip mill. For good rollability and improved formability, the steel has to be coarse grained. Boron addition causes formation of coarse boron nitride and carbide instead of AlN. The boron treatment thus takes care of interstitial elements on one hand while on the other hand; the rise in hardness remains very low during the drawing and forming processes.

In the work by Dixit [75] on softer grade IF steel, it is reported that higher finishing and coiling temperatures play an important role in the achievement of low hardness, coarser grain size and low YS/UTS ratio.

3.1.2.2 Ferritic Rolling

Many cold rollers have found that the ferritic rolled hot rolled material is suitable for thin gauge cold rolling. However, the steel chosen for ferritic rolling requires very low carbon, which necessitates special treatment. Furthermore, there are limitations in the roughing mill in terms of mill power, roll wear, roll life and in the electrical and hydraulic system owing to lowering of rolling temperature.

Hot rolled strip is conventionally finish rolled above the Ar_3 temperature (i.e. in a fully austenitic state) to ensure a final product with a uniform transformed ferrite grain structure after cooling. If the finish temperature is lowered slightly below the Ar_3 temperature, inter-critical hot rolling (i.e. when both austenite and ferrite are present) results in microstructural inhomogeneity, which is detrimental to properties of hot strip [76]. Further, it may be noted that if a significant amount of austenite to ferrite transformation occurs in the finishing stands, the possibility of cobble increases substantially, particularly in case of IF steels. However, if finish rolling is conducted entirely below austenite to ferrite finish temperature (Ar_1) temperature, the microstructure of steel strip is predominantly ferrite during rolling and following recrystallisation of the deformed ferrite, a uniform structure is present in as rolled product.

In a recent work on low carbon (0.03-0.4 wt.% C) by Perry et.al [77], it has been shown that slab reheated at 1200 °C and above, resulted in more sluggish recrystallisation behaviour compared to reheating at 1100 °C. The explanation of this phenomenon has been based on thermodynamic calculation of AlN fine precipitation potential (FPP). FPP has been defined as the volume fraction of precipitate that is expected to come out of solution during hot rolling. Typically, alloys and processing condition that brought about low FPP, found to recrystallised more rapidly and to a near completion stage than high FPP. Further complete recrystallisation was achieved only in sample that was coiled at higher coiling temperature (680 °C) and had reduced AlN besides lower FPP from lower reheating temperature (1100 °C). However, boron addition in low carbon Al-killed steel is expected to result in low FPP of AlN precipitate and can result into fully recrystallised grain even with higher slab reheating temperature.

Humphreys [78] has shown favourable effect of boron on formability in ferritic rolled low carbon steel (0.02-0.04 wt.%C). They have concluded that dynamic strain aging (DSA) behaviour of low carbon steel is modified in the presence of boron. DSA can significantly influence the microstructural development during processing particularly due to suppression of shear band formation. Shear bands help in promoting the development of {111} texture during annealing, the presence of which is necessary to ensure high plastic anisotropy ratio (r_m) value in final product. Therefore, warm rolling of low carbon steel usually results in poor formability. Addition of boron and chromium promoted the formation of shear band at somewhat higher rolling temperatures than that in unalloyed steel

[78]. This could lead to production of warm rolled low carbon steel with improved formability.

In recent patents [70,71], the influence of boron in improving formability and cold reducibility has been demonstrated. The reason for improving cold reducibility has been attributed to lower hardness and lower n value of hot rolled steel. Microstructures and mechanical properties of low carbon boron steel have been compared with those of boron free steel. Boron added steel has been found to possess product attributes suited for improved cold reducibility i.e. lower hardness, lower strain hardening exponent (n), lower yield strength and higher total elongation. Laboratory based heat treatment experiments were carried out and microstructural evolutions have been correlated to hardness. Hardness values for boron added steel have been found to be lower than those of boron free steel irrespective of austenitising temperature and cooling rate.

Patents search [70,71] on the role of boron in improving the formability indicate that hot rolled low carbon steel strip with an excellent press workability is capable of forming smooth pressed surface consisting essentially of 0.12 wt.% carbon, 0.01 wt.% nitrogen, 0.009wt.% boron and other specific optional elements. The ferrite grain size of the hot rolled steel strip is in between 7 – 9 ASTM number. It has been shown that even with high carbon, manganese, silicon and nitrogen, boron addition is helpful in increasing the cold reducibility / formability of hot rolled steel.

Shanmumugam et.al [79, 80] have inferred in their work that superior drawability is related to the modified microstructure as a result of boron addition. The improved properties have been attributed to free boron increasing the hardenabilty by residing at austenite grain boundaries and thereby delaying the onset of transformation to ferrite resulting in microstructure, containing less ferrite and more pearlite than in boron free steel. The hardenability effect also results in the transformation to pearlite taking place at lower temperatures giving a finer pearlite inter lamellar spacing thus promoting higher strength and ductility.

Addition of boron significantly increased the pearlite content (lower ferrite) compared to the conventional plain carbon material. In addition to the conventional lamellar pearlite morphology, there were also numerous areas of degenerated pearlite where the carbide component (cementite) was less continuous and tended towards a more globular morphology. This degenerated pearlite microstructure is not unusual since the pearlite content in the boron steel (boron treated steel has 29 % more pearlite than plain carbon steel) has increased substantially without an increase in the carbon content. Therefore, an effect on the cementite component of pearlite is to be expected. The degree of this change in pearlite morphology will be greater as the carbon content in the steel is reduced. The degenerated pearlite morphology is not observed in boron treated high carbon steel due to availability of higher carbon to retain a fully lamellar structure.

The effect of boron addition on the mechanical properties and strain aging behaviour of medium carbon steel wire rods have been assessed by Bao-Hong Cao et.al [81], wherein it has been reported that addition of up to 63 ppm of boron by weight (i.e. up to B/N atomic ratio of 0.96) continuously improves the elongation to fracture and uniform deformation and is associated by a decrease in tensile strength. This was related to the decrease in the amount of free nitrogen that results from BN precipitation.

Significant decrease in strain aging index (SAI) and in the rate of work hardening accompanied by that addition of boron was noticed which led to improved cold headability of the steel in as-rolled condition. Further, the authors [81] have highlighted the sensitivity of nitrogen scavenging with boron in the post hot rolling thermal history. Slow cooling rates and a BN precipitation treatment resulted in lower yield strength and lower susceptibility to strain aging.

I.3.2 Cold Rolled Steel

Low carbon steel sheets used typically for auto body fabrication are required to have excellent drawability, ductility and non aging properties. Conventionally, these steels have been produced by batch annealing after cold rolling. However, recently various kind of new continuous annealing processes [82-85] have been developed to produce steel sheets having the same quality as in batch annealed products. The scavenging effect of carbide, nitride, oxide and sulphide forming elements on C, N and S need to be applied to the alloy design of appropriate steel for deep drawable continuous annealed sheets.

3.2.1 Continuous Annealed Steel

Study on continuous annealing of low carbon Al-killed steel has been carried out by many researchers [86-90] where the importance of carbon and nitrogen in solution in influencing elongation and r_m value have been highlighted. It has been reported that yield strength of continuous annealed and over aged aluminium steels as well as increase of yield strength after aging and bake hardening are directly related to the amount of soluble carbon and nitrogen in ferrite. Boron is likely to reduce solute carbon and nitrogen in ferrite and influence the morphological changes of carbides, its role in continuous annealed cold rolled steel is important.

Effect of boron addition on the mechanical properties of continuously annealed boron bearing Al-killed steel sheets has been widely studied **[67, 87-90]** wherein the beneficial effect of boron for improving forming properties is highlighted. It has been reported that boron addition increases the total elongation of Al-killed steel sheets by improving the grain growthability by substituting large BN for AlN and that maximum elongation can be attained by optimizing the boron / nitrogen ratio **[67,90,91]**.

The deterioration of elongation, when the boron content is high, has been attributed to grain refinement caused by existence of excess boron or by fine precipitation of $Fe_{23}(CB)_6$ **[66]**. Further it has been shown **[92-93]** that small amount of boron in solution does not cause grain refinement in low carbon steels, and also, the precipitation of $Fe_{23}(CB)_6$ has only been observed in high temperature coiled bands.

The important observation by Hosoya et.al **[89]** shows that carbide morphology, which affects elongation, is changed by boron addition. Thereby it can be considered that change in carbide morphology by excess boron might be the reason for deterioration of elongation.

The most important finding of Funukawa **[87]** is the effect of boron on decreasing strain-hardening value (n). Fig. 11 shows the effect of carbide diameter on n value of boron bearing Al-killed steel. When the carbide diameter is less than 0.5 micron, n value decreased drastically.

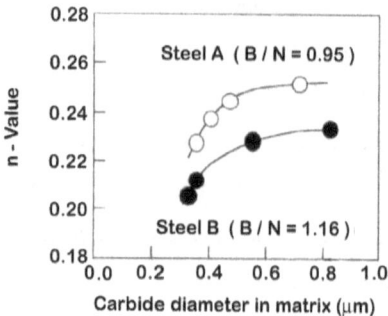

Fig. 11 Effect of carbide diameter on n-value of boron added Al-killed steel sheet

The work hardening exponent has generally been used for the estimation of uniform elongation. A clear correlation has been found between the carbon in the solution and the n value [89]; nevertheless, the decrease in the n value is considered to be caused by the morphological and qualitative changes in carbide. Further, the authors have shown that relatively high work hardening rate in the low strain region, as studied by MF Ashby [94] also. Here the movement of the primary glide dislocation is impeded by the highly dense secondary dislocation forests that are multiplied in the vicinity of second phase particles. In the high strain region, series of sub structural changes in steel with fine carbides are considered to be the fundamental cause which decreases n value.

In other study on continuous annealed cold rolled steel by Takahashi et.al [88], it has been shown that deep drawability and ductility development in low carbon steel is due to sufficient growth of grains in steel during annealing which is attempted by the scavenging effect of alloying elements on impurity atoms. In this study, the content of alloying elements such as manganese, aluminium, boron or titanium within the stoichiometric composition range of precipitates formed by the reaction between alloying elements and carbon, nitrogen, oxygen and sulphur have been found to play an important role in achieving softer cold rolled products [88]. Because, the presence of fine carbides and carbon in solution leads to decrease in uniform and total elongation, it is important to reduce carbide distribution density as well as solute carbon content during over-aging which can easily be obtained by addition of boron. The effect of boron on r_m value of continuously annealed Al-killed steel sheets has been studied by Takahashi et.al [95] .They have observed that at optimum B/N ratio, mechanical properties are almost independent of nitrogen content. But with higher B/N ratio (> 1), the hardening rate is rapid in steel with high nitrogen content.

Fig. 12 shows the schematic expression of changes in properties of boron added steel with carbon and nitrogen. Fig. 13 shows the relationship between excess boron and mechanical properties. It is evident that with increase in excess boron, formability in terms of YS, UTS and elongation decreases; and whereas ideal condition of zero value of excess boron imparts best forming properties.

Further it has been reported [95] that boron levels equivalent to nitrogen (B/N 0.8 to 1.0) can develop high r_m value in case of continuous annealing. The reason for this is related mainly to extreme grain growth after recrystallisation as a consequence of boron addition. Increasing B/N ratio to more than one deteriorates r_m value.

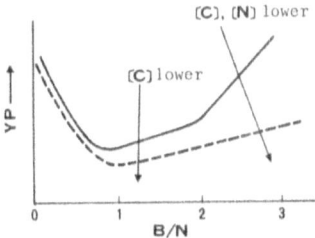

Fig. 12   Schematic depiction variation of yield strength (yield point) with B/N ratio in a B-added steel

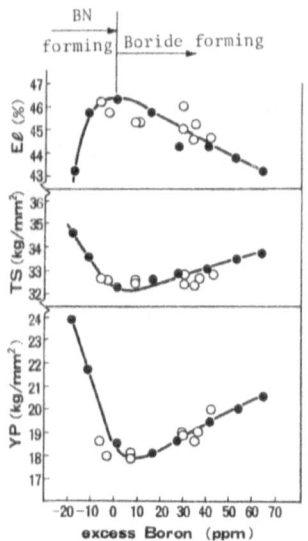

Fig. 13   Relationship between excess B and mechanical properties

Pradhan [67] has further substantiated the above results. For the boron added steel the author has made an important observation that with respect to effect of temperatures of hot rolling, coiling and continuous annealing, the optimum combination of low YS and high r_m value was achieved on stochiometric addition of B which corresponds to B/N ratio of 0.8. YS and r_m value have been explained based on precipitation behaviour of the respective nitrides / sulphides / carbides and the resultant ferrite grain size.

It is underlined that varying coiling temperature (565 – 675 °C) or the annealing temperature (675-850 °C) has very little effect on these properties. Decreasing the steel carbon content, however, lowered strength and improved r_m value **[67]**.

Messien & Leroy **[66]** have studied the effect of coiling temperature on the properties of boron added low C steel. They have observed that due to extensive

BN precipitation before coiling, nitrogen in solution is reduced to a very low level even while coiling at 650 °C. Reduced Al content proved to be effective in grain coarsening and low aluminum boron steel appeared to be most attractive to produce softer HR product.

3.2.2 Batch Annealed Steel

Although boron has beneficial effect on the forming properties of continuously cold rolled (CR) steel, only a few work **[96,97]** has been carried out on its role in batch annealed CR steel. It is well established that the superior drawing properties of commercial Al-killed steel derives from proper development of texture during recrystallisation, which in turn depends critically on the control of AlN precipitation during processing.

The basic requirement is to retain aluminium and nitrogen in solution after hot deformation and prior to cold reduction so that these elements are available for "precipitation clustering" at sub-grains or prior cold worked grain boundaries during subsequent annealing. A number of studies have shown that if AlN is precipitated prior to cold reduction, {111} texture development is prevented, and as a result drawability properties of the steel are reduced to that of rimmed steel.

Quinto et.al **[96]** carried out extensive study on the role of boron in low carbon (0.07 wt.%) unalloyed steel processed through batch annealing route. They have observed that addition of more than 15 ppm of boron to Al-killed drawing quality steel reduces its drawing properties to that essentially of rimmed steel. The change of r_m value from 1.7 to 1.2 is associated with characteristic change from elongated to equiaxed ferrite grains, an inhibition of aluminium nitride precipitation in annealed sheet and lowering of intensity of {111} poles. They have reported that AlN precipitation, which is critical to the development of texture, is inhibited by boron additions to such an extent as to nullify its effect in aluminium killed steel. However, the mechanism by which boron acts to retard AlN precipitation has not been explained.

Precipitation of AlN can be affected by changes in relative activities of solute elements involved, by competing with other precipitation favoured by thermodynamic or kinetic factors, which govern the ease with which certain precipitates can be nucleated. Since boron modifies cementite precipitation, it is interesting to observe that carbon percentage has an effect on AlN precipitation. Ichiama et.al [97] have shown that increasing carbon levels from 0.002 to 0.01 in Fe-Al-N alloys accelerates AlN precipitation. This is presumably because the coexisting carbon atoms reduce solubility of nitrogen and the fine cementite particles furnish more sites of AlN precipitation.

Further, in their study, Ichiama et.al [97] have shown that in presence of 15 ppm of boron (Fig. 14), the available nitrogen is approximately 0.003 wt.% to precipitate as AlN, which is above the minimum nitrogen required for high r_m value. Therefore, non-availability of nitrogen in solution to combine with aluminium cannot be the reason for low r_m value of steel.

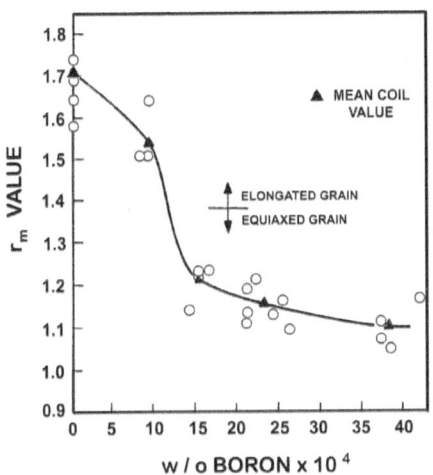

Fig 14 Effect of B addition on r_m value of commercial Al-killed steel

Low r_m has been related to suppression of AlN precipitation by reducing the AlN nucleation sites with fine carbides in presence of boron. Presence of segregated boron or coarse borocarbides act to reduce free energy of these boundaries. Reduced number of effective sites ultimately suppresses AlN-precipitation. A secondary cause for poor {111} and {211} texture-components is also due to effect of coarse carbide during recrystallisation in a boron added steel. The suppression of AlN precipitation and induced precipitation of coarse carbide by B addition in Al-killed steel, singly or in combination can therefore result in a final texture that is essentially that of low r_m typical of rimmed steel. The author has concluded that in the presence of boron > 15 ppm, r_m value of batch annealed aluminium would always be inferior and equal to that of rimming grade (r_m : ~ 1.2), and also the microstructure of final product would be equiaxed with no pancaking, which is typically observed in Al-killed steel.

In a recent study [98], it has been reported that the non-availability of nitrogen to precipitate as AlN is the principal reason for low r_m value in Al-killed batch annealed steel which is not in line with the work of Quinto et.al [96]. Effect of B/N atomic ratio on the forming behavior of low carbon batch annealed aluminum killed steel has clearly demonstrated [98] that B/N ratio can be optimized. Sufficient nitrogen is available in solution to combine with aluminium during batch annealing so that lower YS, higher elongation and almost similar r_m value are obtained in boron added steel which is similar to those observed in boron free steel. Advantage of this study can be exploited in development of cold rolled batch annealed formable steels even with higher nitrogen content.

3.2.3 IF steel

Over the last two decades, ultra low carbon interstitial free (IF) steel sheets have been widely applied to automobile parts because of the excellent formability [100,104]. It is said that recrystallisation texture favourable for deep drawing develops in IF steel because Ti or / and Nb scavenges interstitial solute atoms.

It is also well known that brittle fracture, or the so-called "Secondary Cold Work Embrittlement (SCWE)" occurs by impact loading at a low temperature after press forming in IF steel. This phenomenon is attributed to weakening of the grain boundary by depletion of interstitial atoms at grain boundary. Kimura [99] reported that intergranular fracture was suppressed by carbon addition. Machara and Mizui [100] reported that the coarse ferrite grain structure and transgranular work hardening by press forming accelerates SCWE in IF steel. It is known that SCWE is suppressed by the addition of a small amount of boron.

Suda et.al [101] showed that the ductile brittle transition temperature (DBTT) was lowered by boron addition. Two mechanisms have been proposed to explain the suppression effect of boron on SCWE. One is the strengthening of grain boundary by B-segregation and the other is the decrease of P-segregation, which promotes embrittlement; as a result of site competition behaviour of B and P atoms. Yasuhara et.al **[102,103]** reported that both mechanisms contributed to the suppression of SCWE. However, some studies **[104,105]** have reported that recrystallisation temperature rises and r_m value of steel sheets decreases with B addition. The authors presumed that retardation of TiC (titanium carbide) precipitation, recrystallisation and grain growth would be the cause of the deterioration in r_m value. Haga et.al **[106]** in their study have shown that effect of B on recovery and recrystallisation is strongly related to B-segregation to grain boundary.

References

1. Pickering F.B., Physical Metallurgy and the design of steel, Applied Science Publisher, London, 1978, p.50.

2. Jonas J. J and Akben M. G., Metall. Forum, 1981, Vol.4, p.92.

3. Kwon O. and DeArdo A.J., Acta Metall. Mater, 1991, Vol.39, p.529.

4. Grange R.A. and Garvey T.M., Trans ASM, 1946, Vol. 37, p.136.

5. Grange R.A. and Mickel J.B.,Trans ASM, 1961, Vol. 53, p.157.

6. Morral J.E. and Cameron T. B., Conference proceeding on Boron in Steel, edited by Banerjee S. K. and Morral J. E., The Metallurgical Society of AIME, Warrendale, PA, 1979, p.19.

7. Ohmori Y. and Yamanaka K., Conference proceeding on Boron in Steel, edited by Banerjee S. K. and Morral J. E., The Metallurgical Society of AIME, Warrendale, PA, 1979, p.44.

8. Cameron T.B. and Morral J. E., Conference proceeding on Boron in Steel, edited by Banerjee S. K. and Morral J. E., The Metallurgical Society of AIME, Warrendale, PA, 1979, p.61.

9. Pakrasi S., Just E., Betzold J., Hollrigi-Rosta, Volkswagenwerk AG., Forschung and Entwicklung, Conference proceeding on Boron in Steel,

edited by Banerjee S. K. and Morral J. E., The Metallurgical Society of AIME, Warrendale, PA, 1979, p.147.

10. Yamanaka K. and Ohmari Y., Transaction ISIJ, 1977, Vol.17, p.92.

11. Saikat K De, Anjana Deva, S Mukhopadhyay, B K Jha & S K Chaudhuri, Steel India, 2007, Vol.29, p.61.

12. Deva Anjana, De S. K. and Jha B. K., Material Science and Technology, 2008, Vol.1, p.124.

13. Deva Anjana, De S. K. and Jha B. K., Journal of Materials Engineering and Performance, ASM International, Vol 18, No 1, Feb 09, p. 109.

14. Deva Anjana, Jha B. K. and Mishra N. S., Journal of Material Science, July 2009, Vol 44, No 14, p. 3736.

15. Deva Anjana, Kumar Vinod, De S.K., Jha B.K. and Chaudhuri S.K., Material & Manufacturing Process, 2010, Vol. 25, p.99.

16. Deva Anjana, Jha B. K. and Mishra N. S., Material Science and Engineering A, Vol.528, issue 24, September 2011, p.7375.

17. Deva Anjana, Jha N.K. and Jha B.K, International Journal of Metallurgical Engineering 2011; 1(1)p.1.

18. Grossmann M.A., Trans. TMS-AIME, 1942, Vol. 150, p.227.

19. Corbett R.B. and Williams A.J., U.S. Bur. Mines, Rept. Invest. No. 3816, 1945, p. 21.

20. Crafts W. and Lamont J. L., Trans. TMS-AIME, 1944, Vol. 158, p.157.

21. Udy M.C. and Rosenthal P.C., Trans. TMS-AIME, 1947, Vol. 172, p. 273.

22. Rahrer G.D. and Armstrong C.D., Trans. Am. Soc. Metals, 1948, Vol. 40, p.1099.

23. Rohal L.J., J. Iron Steel Inst., 1954, Vol. 176, p.173.

24. Sakuraya K., Okada H. and Abe F., ISIJ International, 2006, Vol.46, No 11, p.1712.

25. Digges T.G. and Reinhart F.M., Trans. Am. Soc. Metals, 1948, Vol. 40, p.1124.

26. Speight G.E., Discussion, J. Iron Steel Inst., 1950, Vol. 166, p.196.

27. Grange R.A. and Garvey T.M., Trans. Am. Soc. Metals, 1946, Vol. 37, p.136.

28. Kapadia B.M., Brown R.M. and Murphy W.J., Trans. of the Met. Soc. of AIME, 1968, Vol. 242, p.1689.

29. Motock G.T. et.al, Boron, Calcium, Colombium and Zirconium in Iron and Steel, John Wiley and sons, New York, 1957, p. 451.

30. Watanabe S. and Ohtani H., Trans ISIJ, 1983, Vol. 23, p.38.

31. Habu R., Miyata M., Sekino S. and Goda S., Tetsu-to-hagane, 1974, Vol. 60, p.1470.

32. Ueno M. and Inoue T., Trans ISIJ, 1973, Vol.13, p.210.

33. Brown et.al, Met Science, 1978, Vol. 8, p.317.

34. Derge G., Trans. AIME, 1946, Vol. 167, p. 93.

35. Grange R.A., Boron, Calcium, Colombium and Zirconium in Iron and Steel, John Wiley and sons, New York, 1957, p.3.

36. Butler J.F. Trans. of the Met. Soc. of AIME, 1962, Vol. 224, p.84.

37. Watanabe S. and Ohtani, Tetsu to Hagane, 1976, Vol.62, p.1851.

38. Kapadia B.M., Hardenability concepts with application to steel, Met. Soc. of AIME, 1978, p.448.

39. Engl B. and Drewes E.J., Technology of Continuously Annealed Cold-rolled Sheet, Met. Soc. of AIME, 1985, p.123.

40. Keown S.R. and Pickering F.B., Met. Sci., 1977, Vol. 11 (7), p.225.

41. Gladman T. and McIvor, Scavandium Journal Metall, 1972, Vol.1, p.247.

42. Habu R., Miyata M., Tamuka S., and Sekino S., Trans. ISIJ, 1983, Vol. 23, p.176.

43. Leslie W.C., Rickett R.L., Dotson C.L. and Walton C.S., Trans. ASM, 1954, Vol. 46, p.1470.

44. Fountain R.W. and Chipman J., Trans. AIME, 1962, Vol. 224, p.599.

45. Yamanaka K. and Ohmori Y., Trans. ISIJ, 1978, Vol. 18, p.404.

46. Habu R., Miyata M., Sekino S., and Goda S., Trans. ISIJ, 1978, Vol. 18, p.492.

47. Takahashi N., Shibata M., Furuno Y., Hayakawa H., Kakuta K. and Yamamoto K., Metallurgy of Continuous Annealed Steel, AIME, Warrendale, 1982, p.133.

48. Shoenberger L. R., Trans AIME, 1958, Vol.212, p.402.

49. Banks K., Stumpf W. and Tuling A., Materials Science and Engg. A, 2006, Vol. 421, p.307.

50. Shyne J.C. and Morgan E.R., Trans. AIME, 1957, Vol. 209, p.116

51. Gladman T., The Physical Metallurgy of Micro-alloyed Steels, The inst. of Met. Min. and Mining, 1997, p.104.

52. Wilson F.G. and Gladman T., International Materials Review, 1988, Vol. 33 (5), p.221.

53. Gladman T., Proc. R. Soc., 1966, Vol. A294, p.298.

54. Gladman T., Proceeding of first Riso. Symp. on Metals and Material Sc. Denmark, Riso National Lab., 1980, p.183.

55. Bai D. Q, Yue S., Sun W. P. and Jonas J. J., Metallurgical Transactiuons A, 1993, Vol 24 A, p.2151.

56. Bai D.Q., Yue S., Maccagno T. and Jonas J.J., ISIJ international, 1996, Vol. 36 (8), p.1084.

57. Mavropoulos L.T. and Jonas J.J., Canadian Metall. Q., 1988, Vol. 27, p.235.

58. Mavropoulos L.T. and Jonas J.J., Canadian Metall. Q., 1989, Vol. 28, p.159.

59. Djahazi M., He X.L., Jonas J.J. and Sun W.P., Metall. Trans. A, 1992, Vol. 23A, p.2111.

60. Lassraqui A. and Jonas J. J., Metallurgical Transaction A, 1991, Vol.22A, p.151

61. Migaud B., Hot working and forming processes, edited by Sellars C. M. and Davies G. J., Met. Soc., London, 1980, p.67.

62. De S. K., Deva Anjana, Mukhopadhyay S., Mallik S, Verma S and Jha B. K., Steel Times International, April 2009, Vol. 33, No 3, p.53.

63. Bleck W., Kohler K., Meyer L. and Preisendanz C., Thyssen Techn. Ber., 1991, Vol.23(1), p. 43.

64. Muschenborn W., Imlau K.P., Meyer L. and Schriever U., Micro-alloying 95 Conf. Proceeding, 1995, p.35.

65. Cho Y.R. and Kim S.I., Iron and Steel Technology, May 2004, p.46.

66. Messien P. and Leroy V., Steel Research 60, 1989, No. 7, p.320.

67. Pradhan R., Conf. Proceeding on Technology of Continuous Annealed Cold Rolled Sheet Steel, TMS-AIME, Warrandale, PA, 1984, p.185.

68. Guillet A., Sadiqi E Es, Lesperance G. and Hamel F. G., ISIJ International, 1996, Vol.36, p.1190.

69. Tanino M., Nippon Steel Technical Report, June 1983, p.331.

70. Bano Xavier anmd Giraud Christian, Patent on hot rolled steel for deep drawing (Patent No : 5,873,957) dated 23rd February, 1999.

71. Nakazato Y., Nakazawa M., Ohashi N and Yo I., Hot rolled low carbon strip with excellent press workability capable of forming smooth pressed surface and a method of making the same, Patent No523537, dated 13th November 1977.

72. Takahashi N., Foruno Y., Nosaka S., Fukuchi T., Asai T. And Awamoto T., Transaction ISIJ, 1980., Vol.20, No 10, p.B451

73. Takahashi N., Foruno Y. and Hayakaw T., Trans ISIJ, 1981, Vol.21, No.6, p. B286.

74. Watanabe S. Ohtani H. And Kunitake T., Transaction ISIJ, 1983, Vol.23, p.31.

75. Dixit J.K., Sadhu M.C. and Venugopalan T., Tata search, 2003, p.346.

76. Kim J. and Kwon O., Proceeding of THERMEC-88, Tokyo Japan, 1988, p.668.

77. Perry A.C., Thompson S.W. and Speer J.G., 41st Mechanical Working and Steel Processing Conf. Proceeding, ISS, 1999, Vol. 37, p.935.

78. Humphreys A.O., Liu D.S., Toroghinezhad M.R. and Jonas J.J., ISIJ International, 2002, Vol. 42, p.S52.

79. Shanmumugam S., Mishra R. D. K., Mannering T., Panda D. And Jansto, Material Science & Engineering, 2006, Vol.437, Issue 2, p.436.

80. Shanmumugam S., Ramisetti N. K., Mishra R. D. K., Hartmann J. And Jansto S. G, Material Science & Engineering, 2008, Vol.478, Issue 1-2, p.26.

81. Cao B.H., Jonas J.J., Hastings P.R., Nickoletopoulos M., 41st Mechanical Working and Steel Processing Conf. Proceeding, ISS, 1998, Vol. 35, p.783.

82. Toda K., Gondoh H., Takechi H., Abe M., Uehara N. and Komiya K., Transactions of ISIJ, 1975, Vol. 15, p.305.

83. Toda K., Gondoh H., Takechi H., Abe M., Stahl U. Eisen, 1976, Vol. 96, No. 25/26, p.1320.

84. Kubo Dera H., Nak Oka K., Araki K., Watanabe K., Nishi Moto A. and Iwase K., Transactions of ISIJ, 1977, Vol. 17, p.663.

85. Paulus P. and Economopoulos M., Stahl U., Eisen, 1978, Vol. 98, No. 17, p.873.

86. Tate M., Muroga O, Arak K., Inove I., Takahashi M, and Watanabe K., Nippon Kokan Technical Report Overseas No.35, 1982, p.14.

87. Funakawa Y., Inazumi T. and Hosoya Y., ISIJ International, 2001, Vol. 41, No. 8, p.900.

88. Takahashi N., Abe M., Akishu O. and Katoh H., Conf. Proceeding on Metall. of Continuous Annealed Sheet Steel, TMS-AIME, Warrandale, 1982, p.51.

89. Hosoya Y., Kobayashi H., Shimumura T., Matsudo K. and Kurihara K., Conf. Proceeding on Technology of Continuous Annealed Cold Rolled Sheet Steel, TMS-AIME, Warrandale, PA, 1984, p.61.

90. Kawamura K., Otsudo T. And Furukawa, Trans ISIJ, 1976, Vo.16, p.538.

91. Sudo M. and Tsukatani I., Conf. Proceeding on Technology of Continuous Annealed Cold Rolled Sheet Steel, TMS-AIME, Warrandale, PA, 1984, p.203.

92. Shimbashi I., Nagira N., Abe M., Sakurai K., Kitajuma S. And Toshimtsu T., Nippon Steel Technical Report, 1981, No.18, p.37.

93. Watanabe S. and Ohtani H., Tetsu-to-hagane, 1976, Vol. 62, p.14.

94. Ashby M.F., Work Hardenability of Dispersion-Hardened Crystals, Philosophical Magazine, 1966, p.1157.

95. Takahashi N., Furuno Y. and Hayakawa H., ISIJ, 1980, B-286, Lecture No. S1247.

96. Quinto D.T. and Hughes I.F., Metallurgical Transactions, 1976, Vol. 7A, p.165.

97. Ichiyama T., Koizumi M., Yoshida I., Watanabe K. and Nishiyumi S., Transactions ISIJ, 1970, Vol. 10, p.429.

98. Deva Anjana, De S. K. and Jha B. K., Material Science and Technology, 2008, No.1, p.124.

99. Kimura H., Buletin Japan Institute Metallurgy, 1985, Vol. 24, p.376.

100. Machara Y., Mizui N. and Arai M., Proceeding of the International Symposium on Interstitial Free Steel Sheet, Edited by L.E. Collins and D.L. Bargar, CIM, Otawa, 1991, p.135.

101. Suda T., Sakamaki M., Tayama K., Araki K., Wada M. and Kajitani H., Tetsu-to-hagane, 1983, Vol. 69, p.S1365.

102. Yasuhara E., Sakata K., Kato T. and Hashimoto O., ISIJ International, 1994, Vol. 34, p.99.

103. Yasuhara E., Sakata K., Kato T. and Hashimoto O., Physical Metallurgy of IF Steels, ISIJ, Tokyo, 1993, p.223.

104. Hosoya Y., Hashimoto T. and Nishimoto A., Physical Metallurgy of IF Steels, ISIJ, Tokyo, 1993, p.179.

105. Tsukatani I., Tetsu-to-Hagane, 1989, Vol. 75, p.774.

106. Haga J., Mizui N., Nagamichi T and Okamoto A, ISIJ International, 1998, Vol. 38, p.580.

i want morebooks!

Buy your books fast and straightforward online - at one of world's fastest growing online book stores! Environmentally sound due to Print-on-Demand technologies.

Buy your books online at

www.get-morebooks.com

Kaufen Sie Ihre Bücher schnell und unkompliziert online – auf einer der am schnellsten wachsenden Buchhandelsplattformen weltweit! Dank Print-On-Demand umwelt- und ressourcenschonend produziert.

Bücher schneller online kaufen

www.morebooks.de

VDM Verlagsservicegesellschaft mbH
Heinrich-Böcking-Str. 6-8   Telefon: +49 681 3720 174   info@vdm-vsg.de
D - 66121 Saarbrücken   Telefax: +49 681 3720 1749   www.vdm-vsg.de

www.ingramcontent.com/pod-product-compliance
Lightning Source LLC
Chambersburg PA
CBHW031552210526
45464CB00003B/1270